The Conservationist's Book

Practical Conservation Tips for the Home and Outdoors

Lisa Capone

Illustrated by Cady Goldfield

APPALACHIAN MOUNTAIN CLUB BOOKS
BOSTON, MASSACHUSETTS

Cover Illustrations: Cady Goldfield
Text Design: Carol Bast Tyler

Copyright © 1992 Appalachian Mountain Club. No part of this publication may be reproduced or transmitted in any form or by any means, electronic or mechanical, including photocopying and recording, or by any information storage or retrieval system, except as may be expressly permitted by the 1976 Copyright Act or in writing from the Publisher. Requests for permission should be addressed in writing to Appalachian Mountain Club Books, 5 Joy St., Boston, MA 02108.

Distributed by The Talman Company.

Library of Congress Cataloging-in-Publication Data
Capone, Lisa.
  The Conservationworks book : practical conservation tips for the home and outdoors / Lisa Capone : illustrated by Cady Goldfield.
    p. cm.
  "Part of a conservation-education program developed by the Appalachian Mountain Club"—Introd.
  Includes bibliographical references (p. 92–93).
  ISBN 1-878239-11-2 (alk. paper) : $7.95
  1. Environmental protection—Citizen participation. 2. Environmental protection—United States—Citizen participation.
3. Conservation of natural resources—Citizen participation.
4. Conservation of natural resources—United States—Citizen participation. I. Appalachian Mountain Club. II. Title.
TD171.7.C37   1992
363.7'0525—dc20   92-6836
                                                                CIP

The paper used in this publication meets the minimum requirements of the American National Standard for Information Sciences—Permanence of Paper for Printed Library Materials, ANSI Z39.48-1984.∞

Printed in the United States of America.

Printed on recycled paper. ♻

10 9 8 7 6 5 4 3 2 1   92 93 94 95 96 97

To my best friend Will,
and to our children
Sally and Scott,
and to the Earth they'll inherit.
L.C.

To my parents,
and to my teachers
for their infinite patience.
C.G.

Contents

Introduction	*vii*
1. Our Home	9
2. Conservation Inside Your Home	21
3. Recycling	45
4. Conservation Outside Your Home	59
5. Conservation in the Backcountry	73
Appendix	91
Sources	92
About the Author/Illustrator	94
About the AMC	95

Introduction

Welcome to Conservation**works**! This handy guide is part of a conservation-education program developed by the Appalachian Mountain Club. For over a century we've been practicing conservation—promoting the protection and wise use of the mountains, forests, rivers, and open spaces of the Northeast. Now we want you to join us. That's why we created *The Conservationworks Book*.

Many of you know something about conservation already. Perhaps your town has a recycling program. Perhaps you've insulated your house to save energy, or learned to cut down on wasteful packaging when you shop. *The Conservationworks Book* is designed to let you go further—to learn how to protect the planet *wherever* you are: at home, around the yard, in the city, and (of course) in the backcountry wilderness that we call home.

Chapter One starts you on a journey, revealing the fundamentals of how our planet works and some of the modern environmental problems we face. From there you will learn a variety of practical conservation tips, from home water conservation and smart shopping to eco-safe yard work and backcountry wilderness protection. Most of all, you will learn the important contribution *each individual* can make to protect the planet. And we hope that while you and your kids are reading the text, you'll get a laugh or two from Cady Goldfield's humorous illustrations.

The Appalachian Mountain Club has been promoting conservation since its founding in 1876. What started as a small group of scientists and professors exploring and mapping the wilds of New England has been transformed into a major conservation organization. Each year over 65,000 people spend the night in our facilities; over 250,000 people pass through our Pinkham Notch Visitor Center in New Hampshire; and thousands more come into contact with us through hiking, skiing, canoeing, and walking. It was from the need to reach all these people and get them more involved with conservation that Conservation**works** was born.

Under the Conservation**works** program, students from the inner city are rebuilding a local park, AMC hut crews compost food waste in the mountains, eighth graders have developed environmental commercials, and students and teachers together have developed recycling programs in their schools. Whether it's turning off lights, installing efficient bulbs, brushing your teeth with a glass of water instead of a running faucet or putting on a sweater to keep warm, conservation is easy. This book tells you how.

The Conservation**works** program has become a broad-based educational program that reaches our visitors and guests as well as schools, teachers, and the community. Our program uses posters, books, and hands-on workshops to inform and educate. A wide variety of people—naturalists, volunteers, professors, teachers, and youth workers—help us to reach out and make certain Conservation**works** really does work. But we need to involve more people. To make it work we need you.

So read this book. Get out there and get your hands dirty. Put on a sweater and roll up your sleeves. The planet needs you! And once you get involved you'll see why we say: Conservation**works**.

—Walter Graff
Education Director
Appalachian Mountain Club

Note: As you read *The Conservationworks Book,* sometimes you will see an arrow (↑) in the text. This arrow tells you which illustration goes with the passage you're reading. Enjoy!

1 Our Home

All life is a web, woven together. ↑ Look at this forest scene. Anyone can see its beauty, but look closer. The plants, animals, and land fit together into a web that sustains all as long as no thread is broken. Webs like this are everywhere. They form the ecosystems of the world.

This ecosystem is a good example. Mother Raccoon is out for a quick hunt before heading back to her young. At a forest stream she catches a fine fat frog. The frog has in turn eaten worms and snails, which have eaten decaying leaves and acorns. The raccoon's wastes help feed the plants. Even molds and pesky bugs have their place—they help dead vegetation decompose and replenish the soil. Then the soil is ready to feed the plants and trees that provide food and shelter for animals, produce oxygen, and consume carbon dioxide. ↓ This circle of nature's feeding systems is called the trophic cycle.

NUTRIENTS FROM DECOMPOSED ANIMAL WASTES & REMAINS

TROPHIC CYCLE

→ Air, land, and water are also part of the web. In the water cycle, rain replenishes the stream while the sun's warmth evaporates water from it. The water vapor from the stream—and from the forest canopy and even the breath of the animals—eventually finds its way upward in the atmosphere, where it condenses into clouds that rain back on the earth, refilling the stream.

SUN EVAPORATES WATER FROM PONDS, PUDDLES, ETC.
ZOT!
PRECIPITATION FROM CLOUDS
PLANTS & ANIMALS GIVE OFF WATER VAPOR
WATER VAPOR RISES AND FORMS NEW CLOUDS
THE H2O CYCLE
(THAT'S WATER!)

Nature Recycles

When water from the stream rises to form clouds and returns to Earth again as rain, Mother Nature is recycling, something she has done since time began. She keeps her "household" clean and within its budget by wasting nothing and using everything more than once.
↓ You can easily see Mother Nature's recycling at work. The next time you're hiking in the woods, gently kick aside the top layer of debris on the forest floor. What you'll find is a natural compost heap. The top layer of leaf litter is what's fallen to the ground in the past year. Underneath it, you'll find a layer of decayed matter called "duff," that is no longer recognizable as leaves, sticks, and acorns. At the bottom is the completely decomposed rich black humus that makes forest plants thrive. In this way, dead leaves are recycled into something useful to the system that produced them.

LEAF LITTER (LAST YEAR'S STUFF)

DUFF (2 YEARS +)

HUMUS (COMPLETELY DECAYED STUFF)

A RESOURCEFUL WAMPANOAG

Our Ancestors

Plants, animals, and other parts of the natural world don't think about how the pieces of the ecological puzzle fit together. They just instinctively do the right thing. But what about human beings? Unfortunately, too often we've seen ourselves as enemies of nature instead of partners with it.

It hasn't always been this way. Our ancestors lived more frugally and in better harmony with nature because they had to. Those hunters and gatherers would have perished without wisely using the natural resources that sheltered, fed, and clothed them. ↑ Native Americans such as the Wampanoags of New England knew how to conserve resources. For example, if a man broke a knife, he wouldn't just throw the blade away. Instead, he would recycle it into arrowheads or drill points or use it to strike fire.

The European settlers in America were also much better conservationists than we are. A barrel used to transport wine from England to the colony might carry furs back to Britain, more food back to Plymouth, and then end up in someone's home storing corn. Food and other provisions were purchased in bulk and containers to hold them were used until they fell apart. Everyone had a compost, or "muck," heap where they'd toss out pretty much everything, then once a year turn it back into the garden. ↓ And parents' worn-out clothing was often remade into children's garments—another way our ancestors "recycled" what they could.

DON'T WORRY - YOU'LL GROW INTO THEM!

Some Modern Problems
Things have changed a lot in the years since the first European settlers came to America. Our world has become much more complex. The growth of industry has led to better lives for many people, but it's also led to some big problems, too. Here are some:

Global warming
Several scientific reports have shown that the Earth's temperature has warmed an average of .5 to 1.1 degrees F over the past century. Some or all of this warming may be caused by the burning of fossil fuels, which releases so-called greenhouse gases. These gases—carbon dioxide, methane, chlorofluorocarbons (CFCs), and nitrous oxide—trap the sun's heat and prevent it from escaping back into the upper atmosphere. ↓ This "global warming" turns up the heat outdoors. If the trend continues, scientists predict that polar ice caps will melt and ocean levels will rise, inundating coastal areas. Some predict the Earth will be 4° to 9° hotter by the middle of the next century if we don't reduce our use of fossil fuels and CFCs.

Ozone depletion

Stratospheric ozone is a naturally occurring substance that forms a layer in the upper atmosphere. This layer shields the Earth against harmful ultraviolet radiation from the sun. But when chloroflourocarbons—which are used for refrigeration, air-conditioning, and cleaning sprays—are released into the atmosphere, they damage the ozone layer. Another chemical group, halons, used in fire extinguishers, also deplete ozone.
↓ It's hard to believe that such seemingly beneficial things as fire extinguishers and spray cans can harm our ecosystem, but every year more evidence comes to light.

In 1985, a hole the size of the continental U.S. was discovered over Antarctica. Many scientists now think the entire ozone layer is deteriorating quickly, exposing the Earth to damaging levels of ultraviolet radiation.

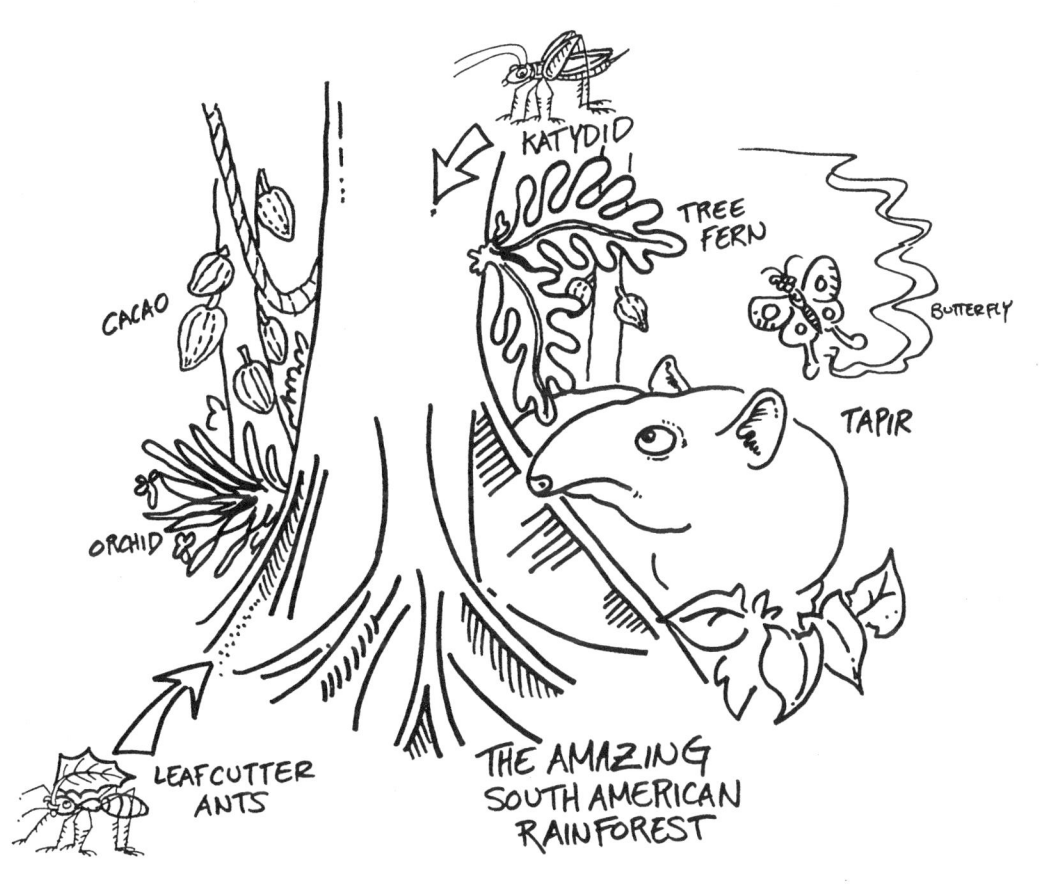

Rain forest destruction
↑ Another environmental problem is destruction of tropical rain forests. Rain forests are natural absorbers of carbon dioxide. They once covered 5 billion acres, but now only half of original rain forest acreage remains, and they are being chopped down and burned at the rate of 100 acres per minute—an area the size of Kansas each year. As they are destroyed, carbon dioxide is released into the atmosphere, increasing the greenhouse effect.

Rain forests harbor approximately half the world's life-forms. Scientists have analyzed only 1 percent of tropical rain forest plants, but those contain more than 25 percent of pharmaceutical compounds sold today. To continue to destroy tropical rain forests is to lose the potential for medical treatments we can't yet envision.

Acid rain

↓ Acid rain is caused when sulfur dioxide and nitrogen oxides from burning coal and oil combine with moisture in the atmosphere to form sulfuric acid and nitric acid. The resulting rain, snow, fog, and other precipitation that returns to Earth is laden with these chemicals. In North America, acid rain has been linked to large-scale damage to trees in Canada and northern New England. Acid rain renders many of North America's lakes and rivers lifeless. In eastern Canada, 14,000 lakes are so acidified that virtually all native fish species are gone.

Waste, toxic and otherwise
↑ Two more pressing environmental problems stem from modern society's wasteful life-style: what to do with the mounting piles of solid and toxic waste? The EPA estimates that Americans produce nearly 160 million tons of ordinary household trash a year, or 3.5 pounds per person daily. That's twice the amount per capita generated in Europe and Japan.

As landfills across the country reach their capacity or threaten public health, communities must decide whether to turn to incineration (with its own host of environmental problems) or source reduction and recycling to solve this problem.

Along with their trash, Americans typically generate 160 pounds each of household hazardous waste annually. This includes stuff like antifreeze, ammonia-based cleaners, paint, drain cleaners, shoe polish, batteries, nail polish and remover, and insecticides. Since it's hard to regulate disposal of these poisonous wastes, they work their way into the "waste stream" and threaten air and water quality.

Consumers aren't the worst offenders, though. American industries release at least 22 billion pounds of toxic wastes into the air, water, and land each year. The General Accounting Office estimates there are between 130,000 and 425,000 potentially hazardous waste sites in the nation. Approximately 1,200 sites are on the EPA's "national priority list" for cleanup, but only 70 sites have been cleaned since the federal "Superfund" law took effect in 1980—over 10 years ago. At that rate, it will be 170 years before these sites get cleaned up!

What to Do?

All these problems may seem overwhelming. But wait! There's lots we can do to fix things. There are ways we can make a difference in our daily lives. That's by practicing conservation, taking advantage of new environmentally sound technologies, recycling, and protecting the outdoors. Let's go on to look at all of them.

2 Conservation Inside Your Home

The simplest and quickest solutions to the Earth's problems are in our own hands. All it amounts to is exchanging some wasteful habits for conservative ones. Change your shopping habits and life-style a little bit ↑—eliminating some of that "easy, use-once, throw away" wasteful stuff—and you will contribute to healing the whole planet. Once you start, you'll find some of these changes don't feel like sacrifices at all. In fact, you'll marvel at your new efficiency. Here are some suggestions for how to make every day Earth Day.

Water Conservation

The average U.S. citizen uses sixty gallons of water each day, and as much as 40 percent is wasted because of habits like letting water run while washing dishes and brushing teeth. Saving water is easy if you follow a few simple habits around the house.

Bathrooms are probably the biggest water-wasters in the house. Over half of household water use occurs there. Conservation measures for the bathroom include:

↓ Shut off the water while brushing your teeth and save up to 10 gallons a day.
• Save 4 to 10 more gallons by turning it off while shaving.

- An electric razor uses no water and only a fraction of the electricity needed to heat water for lather shaving. In addition, the energy needed to manufacture and dispose of blades used in manual shavers is avoided.
- A 5-minute shower instead of a bath saves a whopping 21 to 26 gallons daily, and installation of a low-flow shower head can reduce the flow from 7 gallons per minute to 2.5 without sacrificing comfort.
- If a bath is a must (such as in the case of small children), fill the tub just halfway and save 16 gallons.
- Use faucet aerators to save tap water.

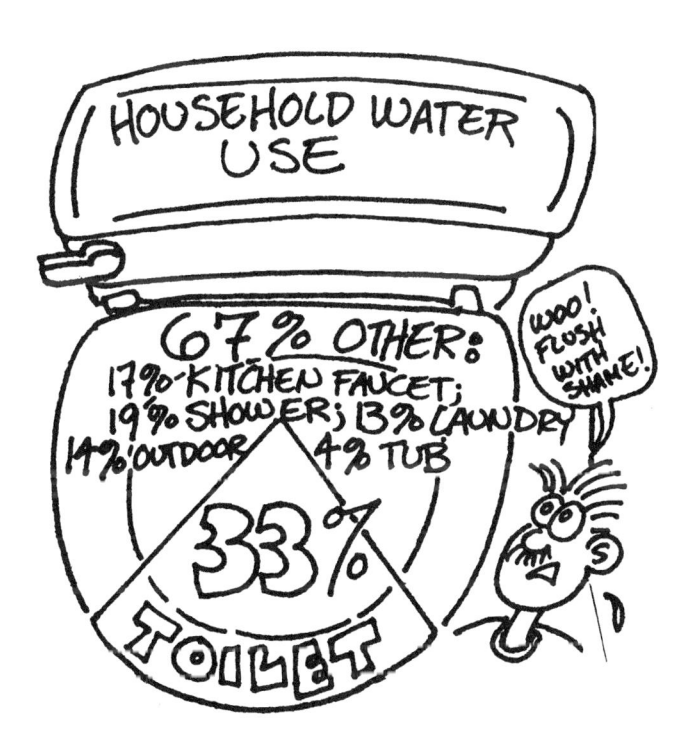

↑ Use low-flush toilets. Fully a third of household water is used to flush toilets, and new models can do the job with as little as 1.5 gallons per flush as opposed to old-fashioned toilets that need 3.5 to 7 gallons. You can install toilet dams or other devices to reduce the amount of water used by older toilets.

- If you can't afford a new toilet or fancy equipment, you can displace some of the water in your toilet's tank with a couple of quart-sized plastic bottles filled with water and a few pebbles. Make sure to sink them out of the way of the toilet's flushing mechanism.
- ↓ Fixing leaky pipes and toilets can also make a big difference. To check for leaks, New England's *Earth Day 1990 Action Guide* recommends that you put about twelve drops of food coloring into the toilet tank. Don't flush and wait fifteen minutes. If the color appears in the toilet bowl, there is a leak.
- The same source says leaky pipes that don't leave a tell-tale puddle can be detected by checking your water meter, recording the reading, and then not using any water for an hour. Recheck the meter and if the numbers are different, you probably need a plumber.

In the kitchen:
- Fill basins for washing and rinsing dishes rather than letting the water run. This can save 8 to 15 gallons per day.
- If the water exiting your tap isn't the right temperature, don't watch it pour down the drain. Keep a container by the sink to catch it and refrigerate it for other uses—drinking, watering plants, cleaning, etc..

In the laundry room:

↑ Use the washing machine only for full loads and save up to 50 gallons per load.

- A properly loaded dishwasher uses less water than washing the same amount of dishes by hand.

Energy Conservation

Home energy use comprises almost a quarter of the country's energy consumption, so you really will be making a difference by making a few changes. You'll also save on heating and electric bills!

Almost half of home energy use is attributable to space heating, and another 14 percent to water heating, according to the Department of Energy (DOE). That makes the basement the best place to start conserving.

- Schedule regular tune-ups for your furnace or boiler.
- If you're thinking of replacing old, inefficient equipment, consider new high-efficiency models. They cost more but can save up to 40 percent on annual fuel bills, which means you'll quickly recoup the difference.
- While you're lurking around in the cellar, lower the temperature of your water heater from the "medium" setting of 140° F to 120° F. You'll cut your water heating bill by 18 percent and protect youngsters from scalds in the bargain.

↑ Wrap both the water heater and pipes with insulation to prevent heat loss.
- If you're going away for longer than a weekend, turn the water heater off. If it's electric, you can install a timer to lower the temperature during sleeping hours.
- If you have installed low-flow shower heads and faucet aerators to conserve water, they will also save energy used to heat water. Wash clothes in cold water for more energy savings (except when washing cloth diapers—the diapers of choice!)

Throughout the house, look for more ways to cut energy use:
- Caulk and weather-strip windows and doors and close fireplace dampers when not in use.
- Don't heat unused rooms. Close them off.
- Set your thermostat no higher than 68° F in winter and, for air conditioner users, no lower than 78° F in summer.

↑ Window fans, awnings, and shade trees planted on the sunny side of the house are all ways to cool off without using air conditioning.
• Turn off lights, television sets, and radios when leaving rooms, and use low-wattage bulbs in areas like closets and hallways.

↓ Switch from traditional incandescent to high-efficiency, compact fluorescent bulbs. They cost more than the old models but, just like high-efficiency heating systems, quickly pay for themselves with energy savings. A 15-watt fluorescent bulb emits about the same light as a 60-watt incandescent and lasts ten times as long. Each bulb you buy will save you thirty dollars over its lifetime and keep 1,000 pounds of carbon dioxide out of the air.

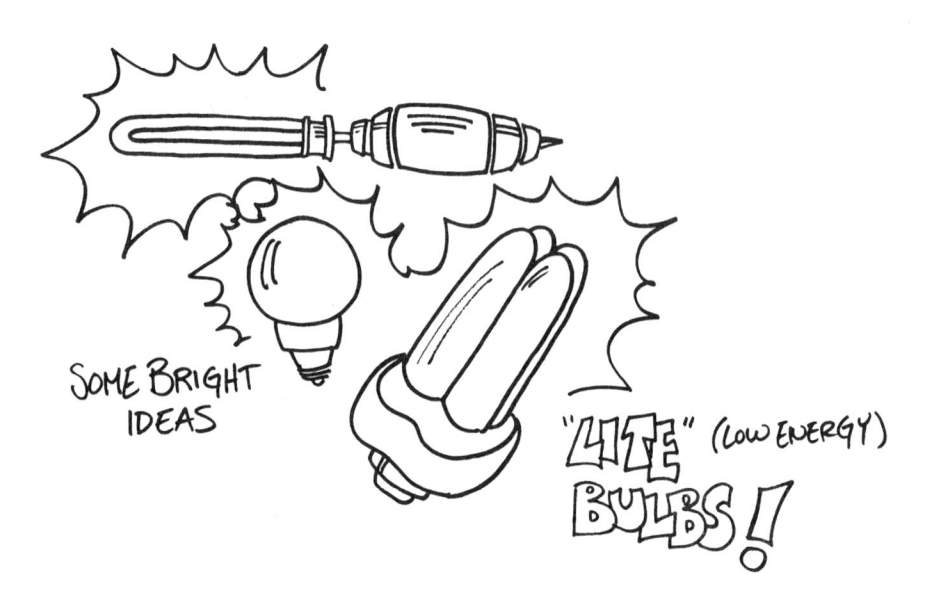

SOME BRIGHT IDEAS

"LITE" (LOW ENERGY) BULBS!

The DOE reports American homes use 13 percent of their energy in refrigeration. To conserve:
- Temperature settings need be no colder than 38° to 42° F for the refrigerator and 0° to 5° F for the freezer. Setting them 10° lower than necessary can increase energy consumption by 25 percent.
- Wash off the refrigerator door gasket to keep it airtight. To quickly check for a leaky gasket, close the refrigerator door on a dollar bill and try to pull it free. If it slips right out without dragging, you have a leaky gasket that's wasting energy.

"NOT SO CLOSE, HOT STUFF!"

↑ Make sure to position the stove and refrigerator away from each other. Dust or vacuum condenser coils behind the refrigerator annually, and keep them at least three inches from the wall.

• If you're buying a new refrigerator, look for freezer-on-top models, which typically use 30 percent less energy than side-by-sides. Cooling fans work more efficiently in a cube-like space than in a long narrow one.

Manufacturers of refrigerators and other major home appliances are required by law to disclose energy efficiency ratings and annual energy costs on yellow Energy Guide labels. Current standards make appliances 10 to 30 percent more efficient than earlier models. Purchasing used appliances is trickier. Although they're cheaper, used appliances will probably cost more to run since efficiency deteriorates with use and since appliances manufactured ten years ago are about 30 percent less energy efficient than current models.

Another way to save energy with appliances is to readjust your use of them. For example:
- Remove clothes from the dryer and hang them up while they're slightly damp. You'll not only save energy from the dryer but save on ironing as well. Many irons run on as much electricity as ten 100-watt incandescent bulbs.
- You can save nearly half the energy needed to operate your dishwasher by opening the door and letting dishes air-dry rather than using the drying cycle.
- Consider alternatives to conventional ovens and ranges. Cook small meals and snacks in a toaster oven, which uses less energy because there's less space to heat. Microwave ovens also conserve energy because of shorter cooking times (but beware of overpackaged foods marketed specifically for microwave users!).

Efficient Transportation

This one's a challenge. America's love affair with the automobile began back with Henry Ford and, by all accounts, the bloom isn't off the rose yet. The Motor Vehicle Manufacturers Association estimates there were 188.6 million cars on the road in the U.S. in 1989. The environmental damage is widespread: gasoline-burning car engines produce several greenhouse gases, most notably carbon dioxide. Also, the disposal of car batteries, tires, and scrap metal contributes to the trash problem. But you don't have to give up your car entirely. Here's how transportation conservation works:
- Tune up your car every 10,000 miles and gain 3 to 9 percent more fuel efficiency. Boost efficiency another 10 percent by keeping tires properly inflated. Keep brakes properly adjusted to prevent fuel-wasting drag.
- Fuel efficiency: new car buyers should shop for a vehicle that gets at least 35 MPG. Buying a highly efficient model saves gas money and encourages more companies to make and market fuel efficient cars.
- Don't keep the car running if you'll be stopped for more than a minute. Try to avoid quick acceleration.

↑ See if you can live without air-conditioning in your car, especially if the weather in your area rarely gets hot for more than a couple of months a year. Car air conditioners are the single biggest users of CFCs—25 percent of U.S. consumption—and thus contribute to ozone depletion when they leak or when the car is junked.

• If you already have a car air conditioner, don't recharge a leaky system yourself with cans of freon gas available at auto supply stores, and make sure your service station recycles CFCs and doesn't just release them to the air. You'll have some help with this soon: federal law requires that all repair shops recycle CFCs by January 1993, and three states—Vermont, Maine and Wisconsin—have passed legislation banning the sale of new cars equipped with CFC-operated air conditioners. A safer alternative to CFCs, HFC 134A, already exists. Let manufacturers know you want them to use it.

• Ensure your safety and protect the environment at the same time by easing up on the gas pedal. Driving 55 MPH instead of 65 reduces the risk of accidents and can increase fuel efficiency by 17 percent.

• Carpool with friends and coworkers to cut down on pollution and traffic.

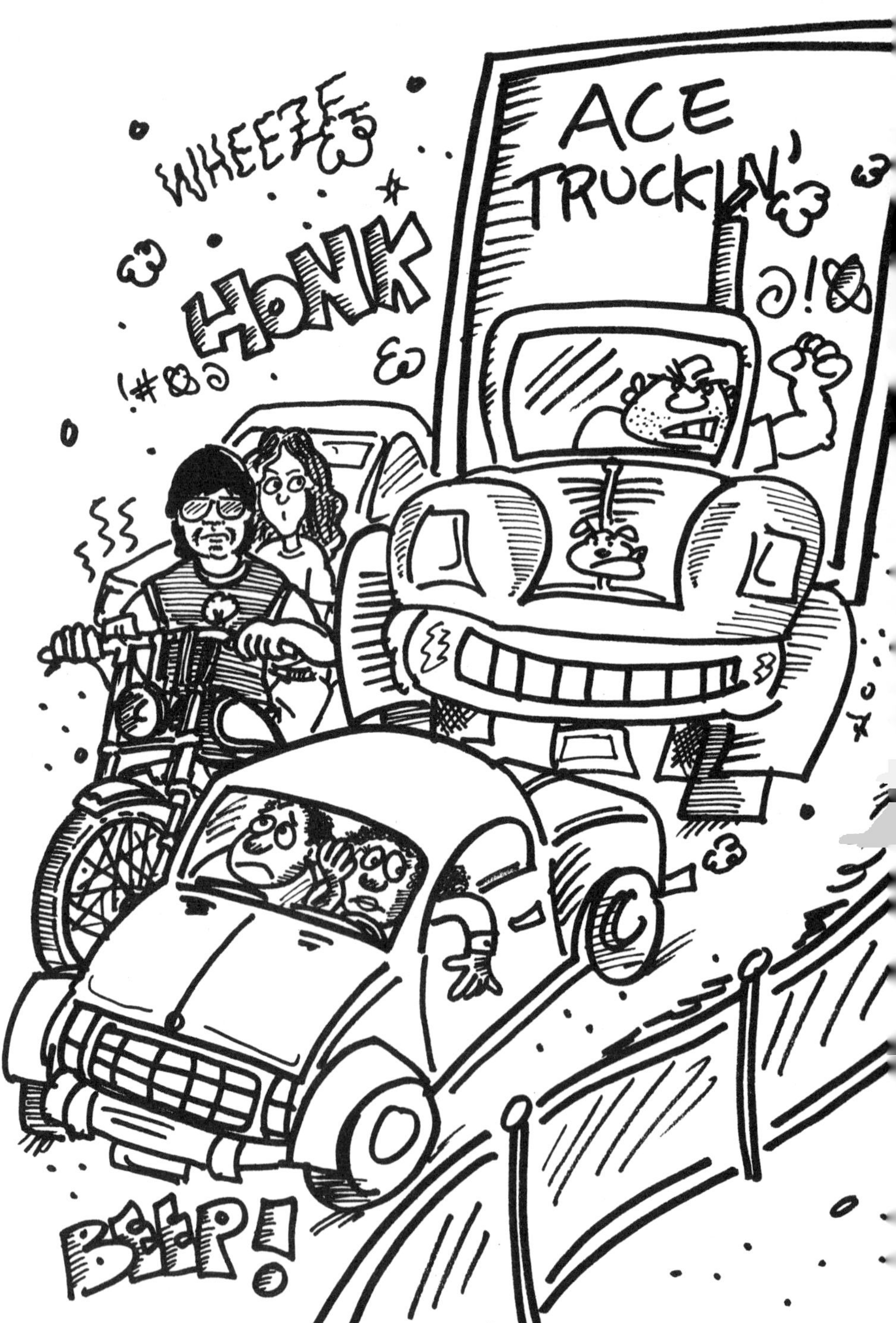
ACE TRUCKIN'
WHEEZE
HONK
BEEP!

← Here's a great way to save energy: challenge yourself to leave your car in the driveway at least one day a week and get where you're going on foot, bicycle, or public transit. According to *50 Simple Things You Can Do to Save the Earth,* if just 1 percent of American car owners did this, each year we would save 42 million gallons of gasoline and prevent 840 million pounds of carbon dioxide from reaching the atmosphere.

Green Shopping

Paper or plastic? That question popped up at check-out aisles across the nation a few years ago when supermarkets responded to consumer backlash over grocers' almost universal switch to plastic shopping bags. Since then there's been considerable debate over which is really better. Paper depends on destruction of trees. Plastic, on the other hand, is made with nonrenewable petroleum and natural gas.

Try to reuse store-provided plastic and paper sacks as many times as you can. On your next return from the supermarket, after putting groceries away, assemble a shopping "kit"—your reusable cloth bags plus several paper or plastic ones and a few clean plastic produce bags—and put it somewhere you won't forget the next time you go shopping. Paper and plastic grocery bags can also be reused as trash bags, eliminating the need to purchase bags specifically made to be thrown away!

↑ Instead of paper or plastic, invest in a few cloth or string bags that can be reused every time you shop.

Some more ways conservation works at the grocer's:
- Select items packaged in recycled materials. Recycled paperboard boxes or plastic bottles often bear the chasing arrows logo.
- Purchase bulk foods such as large sacks of rice, flour, and sugar and family sized containers of peanut butter, jelly, salad dressing, and cereal.
- Look for reusable containers, such as jelly jars that can be reborn as juice glasses or widemouthed glass jars that can be used over and over to store food.
- If your store produce section gives you a choice of taking produce loose or in a plastic bag, skip the bag or use one from your prepacked shopping kit.
- Buy locally grown produce. It takes more fuel to get imported fruits and vegetables to your local store.
- Imports that you should support are nuts and fruits that grow naturally in tropical forests. Produce gathered by rain forest natives gives them a livelihood that preserves rather than destroys the forest. Two examples are Ben & Jerry's Rainforest Crunch candy and ice cream. Part of the proceeds from sales of these products helps fund environmental programs. You can also help stem tropical deforestation by refusing to purchase furniture and other products made from tropical hardwoods.
- Steer clear of hard-to-recycle packaging such as squeezable bottles made from several types of plastics laminated together.
- Avoid juice "brick packs." Brick packs use aseptic packaging—combinations of aluminum and plastic or aluminum and paper—that allow them to be stored without refrigeration. While convenient, aseptic packaging is difficult to recycle. Maine banned aseptic packaging in 1990 and several other states are considering similar action. In response, the Aseptic Packaging Council has launched a multistate program to recycle these special wastes. To find out about starting one in your area, call the council at 800-277-8088.

"PILE O' DIAPERS"

↑ Try to do without disposable diapers. Americans throw away approximately 18 billion single-use diapers annually—enough to reach the moon and back again seven times. Diapers comprise about 2 percent of the solid waste stream and require the paper content of 800 million trees a year. They carry 12,300 tons of human waste and viruses daily into landfills where they can leach into groundwater. After perhaps two hours on a baby's backside the plastic used in disposable diapers takes 500 years to decompose in a landfill, and the average plastic-diapered child will use 7,500 of them before toilet training. Cotton diapers, on the other hand, can be reused up to 200 times before being recycled as rags.

- Whenever possible, purchase durable alternatives to disposable goods such as paper napkins and towels, coffee filters, razors, pens, batteries, and plastic tableware.

MassRecycle of Worcester, Massachusetts, suggests that, as a general rule, green consumers need to ask: "Do I need this item?" "How many times can I use it and how will it affect the environment?" and "Where will it end up after I'm done with it and how will that contribute to pollution?" Such cradle-to-grave thinking helps consumers copy nature's closed systems.

Clean Cleaning

Take stock of your cabinet of cleaning products. Will your house—and the world—be cleaner or more foul after you use them? It's perfectly legitimate to criticize corporate polluters who dump toxic chemicals into the local sewer system, but what about the chemicals we routinely wash down our own household drains?

The good news is that conservation works for spring-cleaning too. Many companies now offer products such as no-phosphate detergents that do less harm to the environment. There are also many tried-and-true traditional cleaners, stuff grandmother probably used before the dizzying array of household chemicals turned up on supermarket shelves. Here are just a few, suggested in both the Gardener's Supply newsletter and the book *The Green Consumer*:

- *Dishwashing detergent.* Soap flakes (such as Ivory), with vinegar for tough grease; in dishwashers, use equal parts borax and washing soda (hydrated sodium carbonate).
- *Toilet cleaner.* Baking soda and a mild detergent.
- *Oven cleaner.* Paste of baking soda, salt, and hot water, or sprinkle with baking soda and scrub with a damp cloth after five minutes (don't let baking soda touch wires or heating elements).
- *Window spray.* First, use alcohol to clean residues left by commercial ammonia-based glass cleaners. Then spray with equal parts white vinegar and water. Dry with cheesecloth or crumpled newspaper.

- *Drain cleaner.* Pour in one-fourth cup baking soda, followed by one-half cup vinegar. Close drain until fizzing stops, then flush with boiling water.
- *Rug cleaner.* Sprinkle baking soda or cornstarch, then vacuum.
- *Furniture and floor polish.* One part lemon juice to two parts vegetable oil.
- *Tub and tile cleaner.* Baking soda and hot water and a firm bristled bursh. Or, rub area with one-half lemon dipped in borax. Rinse and dry.
- *Linoleum floor cleaner.* One-half cup white vinegar mixed into one-half gallon warm water.
- *Silver polish.* Soak in boiling water with baking soda and a piece of aluminum.
- *Air fresheners.* Open a box of baking soda.

High-Tech Help

Traditional cleaners and cloth diapers are useful home remedies for environmental problems, but society must also change on a bigger scale. Today some scientists and engineers are working on high-technology fixes—ways to produce electricity without burning fossil fuels or splitting atoms, natural ways to clean up environmental damage, and so on. Through "appropriate technology," "bioremediation," and other methods, they are seeking to bring high-tech to bear on environmental ills.

- According to the San Francisco-based Political Ecology Group (PEG), a 350-megawatt solar power plant in California now provides electricity at a price competitive with nuclear power.
- Noel Perrin, a professor of environmental studies at Dartmouth College, tries to live a high-tech/environmentalist life. He powers his home with solar panels, and he recently purchased an electric car that will go 70 MPH with recharging needed every 60 to 70 miles. Solar panels installed on the car's roof will mitigate the recharging problem.
- Meanwhile, biotechnology is producing strains of bacteria that have so far been used to clean up oil spills in Alaska and Texas, treat chemical contamination in New York and California, and improve municipal and industrial wastewater treatment facilities.

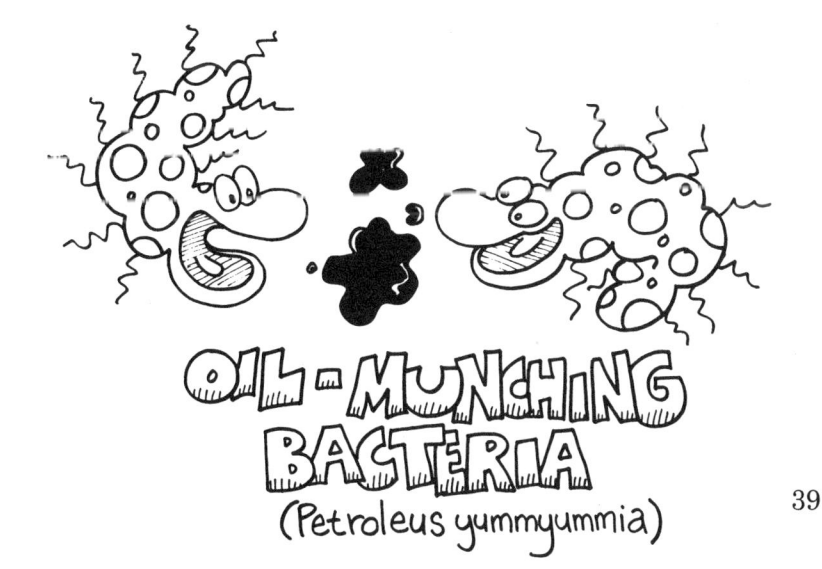

OIL-MUNCHING BACTERIA
(Petroleus yummyummia)

BEST WAY TO RELOCATE A BUG: PUT A JAR OR CUP OVER IT, THEN SLIDE A PIECE OF PAPER UNDER IT. LIFT, CARRY TO OPEN WINDOW, AND RELEASE!

What Kids Can Do

Children are key to making conservation work. Youthful enthusiasm can often prod older family members into making lifestyle changes they otherwise wouldn't. Here are some just-for-kids tips for loving the Earth.

- Respect all forms of life. The next time Mom or Dad is about to squash a bug inside the house, capture the insect and release it outside. Remember, it serves a vital purpose as part of a closed natural system, whether as food for birds or pollinator of flowers.
- Enlist your scout troop, school class or friends in a neighborhood trash cleanup. Make a habit of picking up litter wherever you see it.
- If you take a lunch to school or on a picnic or hike, drink from reusable containers rather than juice boxes or throwaway cups. Use a lunch box instead of a paper bag and reusable containers instead of plastic wrap.
- Go outside to read or do homework on sunny days, rather than turn on electric lights in the house. Use rechargeable batteries in toys and flashlights (you and other family members could chip in to buy your parents a recharger).

↑ Suggest planting trees as part of school science or social studies projects, or help your folks plant some around the yard. They will help combat global warming and provide homes for wildlife.

↓ Make "draft snakes" to put under doors and windows by sewing cloth tubes and filling them with sand, beads, or dried beans.

"DRAFT SNAKES" ARE DRAFT-BLOCKING "COUSINS" TO BEANBAGS...PUT 'EM IN FRONT OF DOORS!

- Sort your family's recyclable paper, glass, plastic, and metal as part of your allowance chores.
- Use your parents' discarded junk mail, old magazines, and other "trash" for arts and crafts projects. It takes creativity to save the world!

What the AMC Is Doing

The Appalachian Mountain Club practices conservation inside its "homes" as well. The AMC runs eight alpine huts for hikers high up in New Hampshire's White Mountains. The huts provide food and shelter, and serve as centers for education and conservation. Conservation is a way of life there, since wasteful habits are impractical in remote mountain settings. Conservation also reminds visitors and staff of the importance of using natural resources wisely. Conserving water in the huts, for example, leaves more outside in mountain streams and helps the animals and plants that depend on them. Here are a few ways the AMC is closing the loop to live more in tune with nature.

↑ The AMC reduced the amount of propane needed to power lamps and refrigerators at Mizpah and Madison huts by installing solar energy systems there in 1989.

- In 1982, the AMC installed water-saver shower heads in its facilities.
- ↓ The huts with flush toilets use low-flow, one-gallon-per-flush models, with an estimated water savings of 2,000 gallons per day. Spring-controlled water faucets at the huts also ensure water conservation.

- The huts are also equipped with compact fluorescent light tubes and emergency radios and smoke detectors powered by solar-charged batteries.
- The AMC's Hiker Shuttle Service encourages low-impact transportation by taking hikers to major trailheads in the White Mountains.

↓ To minimize trash, AMC hut crews buy food in bulk or with minimal packaging. The crews also carry out all non-compostable trash. Visitors to the huts are reminded to carry out whatever they carry in, so there's an incentive to "precycle" and avoid unnecessary products and packaging.

PACKIN' OUT THE RECYCLABLE TRASH FROM THE AMC HUTS

3 Recycling

Recycling is the easiest way to practice conservation at home. It saves natural resources, reduces the amount of trash sent to incinerators and landfills, and conserves energy needed to make new products. It's a simple way to copy closed natural systems.

A typical home "recycling center" uses separate labeled bins for glass, metal, plastic, and paper. Something similar can probably be adapted to your own kitchen, basement, or garage. Some new kitchens are even being built with home recycling in mind: removable bins are built into cabinets so recyclables can be stored and transported quickly and easily.

Sorting

To recycle you need to collect similar things together—green glass with other green glass, newspaper with other newspaper, and so on. This is the first stage of recycling and it's important to do it right because it makes all the other stages much easier to accomplish. Here's how to do it:

↑ Glass: Sort bottles and jars by color—clear, green, and brown. Rinse glass thoroughly and remove labels and plastic or metal neck rings. Remember that the cleaner your containers are the more they're worth to recyclers. Recycled materials compete with well-established sources of virgin materials, so their quality must be high. Recyclables that don't meet the recycler's specifications may end up in landfills or incinerators.

• Paper: Bundle newspapers with string or stack them in brown paper bags—check which your recycler prefers. Some recyclers will take glossy Sunday supplements along with daily papers, but many don't want magazines and phone books because the clay used to produce glossy pages and glue used in bindings complicate the recycling process. See if the phone company will take back your old phone books, and donate magazines to hospitals, nursing homes, schools, or doctors' offices.

↑ Metal: Separate "tin" (really a steel alloy) and steel cans from aluminum ones. You can tell the difference because a magnet will stick to "tin" and steel but not to aluminum. Rinse cans and remove labels. To save space, you can remove the tops and bottoms (recycle them, too) and crush the cans.

Recycled newsprint often becomes paperboard packaging, such as cereal and pasta boxes, but empty paperboard containers may not be accepted by newspaper recyclers. Your community may have a separate program for recycling white "office" paper. If so, you're among the lucky ones who can dispose of junk mail without guilt. If you don't have access to white paper recycling, let the kids use it for coloring, or tear wastepaper into memo-sized sheets and use the backs for phone messages.

• Plastic: There are six kinds of recyclable plastics. They are identified by numbers one through six, usually found on the bottom of containers, often inside the chasing arrows recycling logo. Remember to wash containers, remove lids, and separate out each type of plastic. The most commonly recycled types are numbers one, two, and six.

↑ Plastic type "number one" is polyethylene terephthalate (PET) and is used for soft drink bottles. It gets recycled into new bottles, as well as fiberfill used in jackets and sleeping bags.

• Milk jugs, detergent bottles, and motor oil containers are made of number two, high-density polyethylene (HDPE). Recycling turns them into lumber substitutes, flower pots, pipes, toys, and trash bags.

• Polystyrene foam, known by the trade name Styrofoam, is number six. It comprises fast-food containers, school cafeteria trays, and disposable coffee cups and is recycled into products such as flower pots, hair combs, rulers, and wall and building insulation.

• Numbers three, four, and five—polyvinyl chloride, low-density polyethylene (LDPE), and polypropylene—aren't currently being recycled in any significant volume. Some supermarkets offer to recycle customers' LDPE produce bags and grocery bags.

The field of plastics recycling is still evolving. While recycling a plastic bottle is a lot better than throwing it away (plastic packaging is 10 percent of the solid waste stream by volume), recycled plastic often becomes a different sort of product than what it started out as. This means that another virgin container will replace a recycled bottle or jar. When possible, choose glass over plastic. Glass can be recycled endlessly into the same use it served before.

From Your Home to the Recycler

There are two ways recyclables get from your home to the recycling plant. Either you take them to a drop-off site or waste management contractors pick up your bins at curbside. If you have a curbside program, it's as easy as taking your bins to the sidewalk and watching the recycling truck unload them.

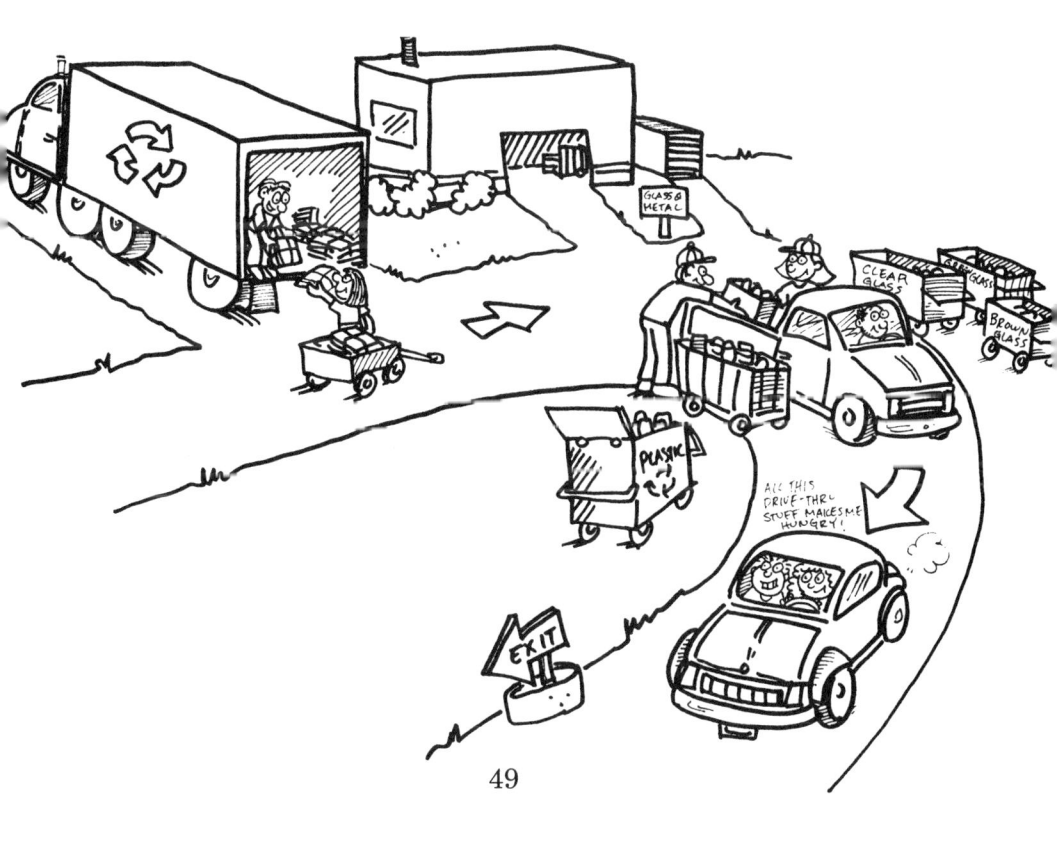

↓ With a drop-off program, you take your recyclables to a local recycling center yourself. After the stuff is collected, recycling trucks transport recyclable materials to a central depot, where they are sorted and consolidated before heading for glass, paper, metal, and plastic manufacturing factories. Businesses and organizations do it a bit differently: they usually contract with commercial collection companies that pick up large quantities of recyclable wastes such as cardboard and white paper.

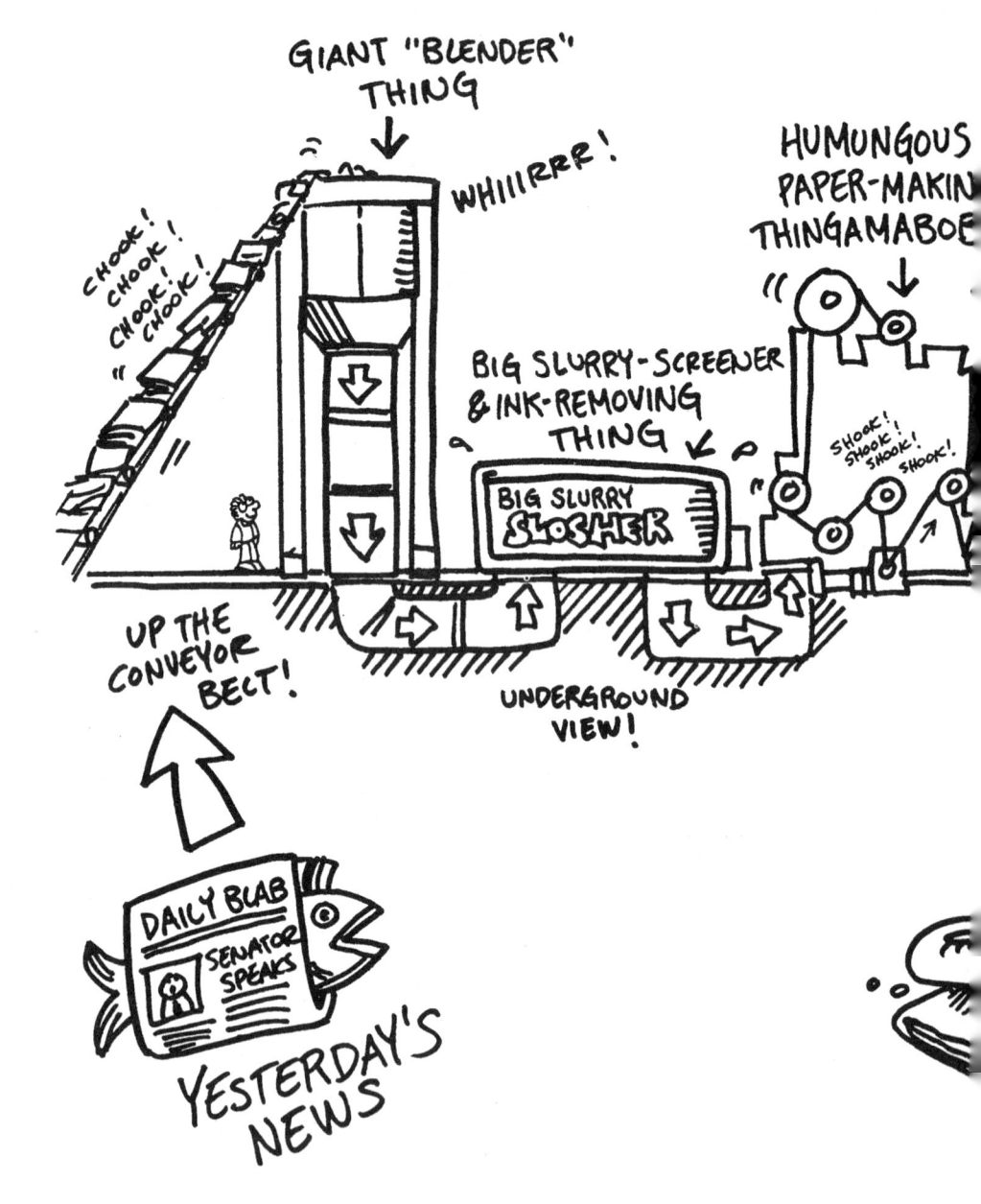

Paper Recycling

← Here is an example of what happens at a paper recycling plant: after it is unloaded at the mill, wastepaper travels by conveyor belt to a pulper, where it is immersed in water and torn apart by rotating steel blades. Chemicals are added to dissolve ink, then the mixture of water and paper fibers, called "slurry," is screened for contaminants (such as staples and paper clips) and washed. Machines press water out of the slurry to produce a sheet of paper. Continuing along the conveyor, the paper is dried, smoothed between rollers, and wound onto large reels. It's then cut into various sizes and may be wound onto cardboard tubes, then loaded onto delivery trucks that transport the recycled paper to its new destination.

The process for glass recycling is similar. At the plant, recycled glass is crushed into small pieces called cullet and mixed with sand, soda ash, and limestone in large furnaces. Molten glass is poured into molds to form new containers, then cooled. The new bottles get loaded on trucks and delivered to other factories where they are filled with foods and beverages.

Purchasing Recycled Goods

Once you've washed the jars, sorted them, and left them by the curb, you're a bona fide recycler, right? Well, almost. Don't forget, nature's systems are closed and recycling has to mimic them to be successful. To close the loop you must buy recycled products. Otherwise, all you've got are nicely sorted piles of trash. ↓ One way to start is to look for recycled packaging in grocery stores. More and more carry products with recycled packaging.

There's a lot of hype about recycled goods being inferior to and more expensive than their virgin counterparts. If that was true in the past, it's not anymore. Many recycled goods, copy paper for example, are competitive in both price and quality with nonrecycled alternatives. Go out of your way to buy them. As the demand increases, prices will continue to fall and more and more stores will carry them. Look for the chasing arrows logo to identify recycled products, but be aware that the logo is sometimes accompanied by the words "recyclable" or "please recycle," indicating that the product can be recycled but is not necessarily made of recycled materials.

↑ Another way to purchase recycled goods is to shop at secondhand stores, thrift shops, and yard sales. Someone else's castoffs may be just what you were looking for. Also remember these outlets, as well as charities, when searching for places to get rid of your unwanted clothing, furniture, outgrown baby equipment, and books.

Creative Recycling

↑ Returning extra clothes hangers to the dry cleaner and sharing seldom-used items such as slide projectors, pasta makers, paints, and rug cleaners with neighbors are more ways to reduce consumption and cut down on trash.

You can also recycle by returning used motor oil to service stations or retailers. Used oil can be cleaned and re-refined as a lubricant or recycled for use as a fuel or fuel supplement. This keeps oil out of the waste stream.

Ceramics, mirrors, and window glass have a different composition from container glass and recyclers won't accept them. Consider the local thrift shop for unwanted dishes and ceramics. Charities will sometimes accept used eyeglasses.

Fun Facts

People sometimes don't realize what a small step like recycling can do to ease environmental problems. Drum up support for recycling in your community by recycling some of these fun facts:

- Recycling a glass jar saves enough energy to light a 100-watt light bulb for four hours (or an energy-saving fluorescent bulb for 16 hours).
- Recycling crushed glass into a new bottle consumes half the water needed to make a new nonrecycled bottle.
- The Wall of Garbage at the Browning-Ferris garbage museum in San Jose, California, is 20 feet high and 100 feet long. It equals the amount of garbage discarded in the U.S. every second.
- The average American uses 7 trees a year in paper, wood, and other products.
- Americans generate enough trash each year to fill a convoy of 10-ton garbage trucks 145,000 miles long. This convoy would circle the equator 6 times or reach halfway to the moon. It's enough garbage—160 million tons—to fill the New Orleans Superdome from top to bottom twice a day, every day of the year.

- 75,000 trees go into making each Sunday edition of the *New York Times*.
- Only 30 percent of newspapers in the U.S. are recycled.
- It takes 17 trees to make a ton of nonrecycled paper.
- Making paper from recycled fibers instead of trees uses half the energy and creates 74 percent less air pollution and 35 percent less water pollution.
- Recycling "tin" (really steel) cans instead of using virgin resources reduces energy use by 74 percent, air pollution by 85 percent, and water pollution by 76 percent.
- Every year, Americans throw away enough steel to reconstruct Manhattan.
- Every three months, we discard enough aluminum cans to rebuild the country's entire commercial air fleet.
- Americans throw away 2.5 million plastic bottles every hour.

↑ According to *50 Simple Things You Can Do to Save the Earth,* 26 plastic soda bottles can be recycled to make one polyester suit.

↓ At-home mechanics dump 180 million gallons of motor oil on the ground, down a storm drain, or in the trash each year. That's 16 times the size of the Exxon *Valdez* oil spill and 5 days' worth of heating oil for the entire U.S.

PACKING TRASH OUT AT PINKHAM NOTCH, NH

How the AMC Recycles

↑ The Appalachian Mountain Club gets its guests to recycle their trash by not providing wastebaskets in its rooms at the Joe Dodge Lodge, its hostel facility at Pinkham Notch Visitor Center, at the foot of Mt. Washington in New Hampshire. Instead, visitors are expected to use the recycling centers at the lodge and in the trading post. There are bins for glass, PET and HDPE plastics, tin, metal, and paper. Pinkham Notch also recycles computer paper, office paper, and newspapers, as well as cardboard and batteries. Hut-generated trash is packed down to Pinkham for transport to various recycling centers. By recycling and composting waste, the hut system reduced its trash by 15 percent in 1990. By 1992, the hut system hopes to have reduced trash tonnage by 90 percent. The AMC's Boston headquarters recycles office paper, beverage cans, and bottles.

The AMC closes the loop whenever it can. The Joy Street headquarters purchases 100 percent recycled toilet paper, hand towels, and copy paper and 50 percent recycled-content computer paper. Fund-raising and membership mailings go out on recycled paper whenever possible. In addition, the club uses envelopes with either empty or waxed paper address windows, instead of plastic ones that render envelopes unrecyclable. The AMC's monthly magazine, *Appalachia Bulletin*, has been printed on 50 percent recycled-content paper, including the coated cover stock, since October 1990. The books department prints its volumes (including this one!) on recycled paper whenever possible.

4 Conservation Outside Your Home

Conservation inside your home is great, but it doesn't make much sense to practice good habits inside if you hire a lawn care company to soak your grass in weed killer to get that "perfect" suburban lawn. Protecting the land around your home is an ideal way to think globally and act locally: So . . . ↑ Go ahead and write to the nasty chemical manufacturer that's poisoning lakes and rivers, but while you're at it, take stock of your own behavior in your little corner of Earth.

"GONE BUT NOT FORGOTTEN"

Conservation in the Suburbs

Suburban yards typically absorb five to ten pounds of pesticides per acre annually—that's more than gets applied to most farms. Most common lawn pesticides have not been fully tested for long-term safety, and some (↑ such as DDT) that were once thought to be safe have been banned. Many of these chemicals easily travel into aquifers, streams, and ponds, where they pollute water and poison wildlife. They also create imbalance, producing a quick flush of green followed by decline and the need for more chemicals. They are an invisible threat to the web of life around your house.

By learning a few natural lawn and garden care techniques, you can have a lovely yard. Let's begin with the lawn:

- First, a healthy lawn requires at least five inches of nutrient-rich topsoil. If you haven't got it, consider adding composted manure across the existing lawn.
- Grass also needs well-aerated soil. The best natural aerators are ants and earthworms. Ironically, they are also often the first to fall victim to chemical pesticides.
- Add natural nutrients such as organic compost, bonemeal, cottonseed meal, and dried blood.
- Once you've got healthy soil, choose a variety of grass seeds suitable to your region. A garden center can suggest what to buy.

↑ Don't use pesticides on your lawn. There are a couple of natural cures for the most common lawn pests. The white grub of the Japanese beetle will succumb to a commercially available bacteria called milky spore disease (MSD). Cinch bugs can be handled by spraying insecticidal soap over the grass thatch just below the soil's surface.

• Use a soaker hose or drip irrigation system, rather than an overhead sprinkler. Water the lawn frequently until it's established. After that, it needs only about an inch of water (or rainfall) every week or so. It's better to give the lawn one good soak than several light waterings since light frequent watering keeps root systems too close to the surface of the soil. Water early in the morning so moisture gets a chance to seep in before the heat of the day evaporates it.

• Mow the grass with a sharp blade and don't cut it too short (like a golf course). A dull blade will tear grass, causing it to lose moisture, and grass cut too short will be stressed by heat and drought.

• Pull weeds up by hand, in the spring, so you'll get their roots. Fewer will grow back later in the season.

If you do all this and still can't maintain a decent lawn, maybe nature didn't mean for grass to grow in your yard. Why worry about it when there are attractive alternatives?

↑ A meadow garden is a practically maintenance-free choice. Garden centers can supply wildflower mixtures that will do well in your area. Prepare the planting site as you would for a lawn. Once the plants come up, sit back and watch while birds and butterflies flit among your poppies.

Every garden suffers from insect pests now and then. Look for pest-resistant flowers and vegetables and keep them healthy. Remember, too, that your garden can put up with some bugs. When controls are called for, try these home remedies:

- Get rid of aphids, earwigs, tent caterpillars, leafhoppers, and others with a spray of plain or soapy water.
- Put collars made with three-by-five-inch file cards or sawed-off bottoms of plastic bottles around vegetable stalks to ward off cutworms.

- Pick off caterpillars and sawfly larvae by hand.
- Set out stale beer at night to attract and trap slugs.
- Plant marigolds to repel nematodes.
- Try commercially available insect traps designed for specific pests.
- Make your own bug sprays. One whole garlic bulb liquefied with a quart of water in a blender and added to a tablespoon of soap is effective against stinkbugs. If you're not queasy, try collecting a bunch of dead insects and mixing them with water in a blender (not used for food preparation!) or with a mortar and pestle. To one-half cup bugs, add two cups of water and liquefy. Strain the "bug juice" through cheesecloth, cut with four to eight parts water,

↑ Mix two tablespoons of hot red pepper in a quart of water and you've got a repellent for tomato hornworm.

and pour into a sprayer. Treat both sides of the foliage and repeat after a rain. Leftovers can be frozen until the next growing season. No one knows exactly why this works, but some speculate the spray may broadcast diseases present in some of the ground-up bugs.

- Use netting. Keep birds away from ripening strawberry plants by covering plants with a gauze sheet. Gauze can also protect cucumbers, squash, and pumpkins from squash beetles. A simple chicken wire fence will discourage rabbits from dining on your lettuce, as will a few open empty wine bottles buried halfway in the soil. The noise the wind makes when it blows over the bottles frightens small animals away.

↑ Purchase beneficial insects from garden supply dealers. Many insects, including praying mantises, ladybugs, and some wasps, are natural predators of garden pests.

Composting

No matter what you're growing, a backyard compost heap is the ideal way to make your own fertilizer and reduce the amount of garbage you throw out. The simplest method involves nothing more than piling leaves and grass clippings in a corner of the yard and periodically turning them over with a pitchfork (turning the pile accelerates decomposition). For a more actively managed pile, you'll need at least one bin constructed of wire or wood that allows air, water, and sunlight to circulate.

↑ Here's one method endorsed by New England Earth Day in its *1990 Action Guide:* Begin with a layer of sticks, twigs, and other coarse materials that will allow air to circulate under the pile. Add a second layer of dry leaves, weeds, and grass clippings. Top that with green vegetation and kitchen wastes (keep a covered container beside the sink to collect food scraps), and finish with a thin layer of leaves or soil. Keep layering kitchen waste, dry material, and soil until the pile is about three feet high. Then start a new pile while the first one decomposes. Turn the pile occasionally with a pitchfork, and it should reduce to half its size in six months to a year. Compost is ready to use when its color is brown to black, it has no unpleasant odor, and contains no readily identifiable components.

Wildlife

↓ Observing birds, squirrels, chipmunks, and other small animals is one of the great advantages of suburban living. Imagine what might happen with a few landscaping changes to make the local critters more comfortable:

- Food can be in the form of fruit- and seed-bearing plants or feeders that you fill yourself. Plants that are native to your area will attract the most wildlife. Nectar-filled feeders for hummingbirds should be taken down around Labor Day so hummers are encouraged to begin their long journey south before cool weather sets in. Brightly colored (especially red) flowers will attract hummingbirds and butterflies.
- Provide water in a birdbath or fountain for year-round drinking and bathing. A continuous slow drip in a shallow basin will keep water open and clean.

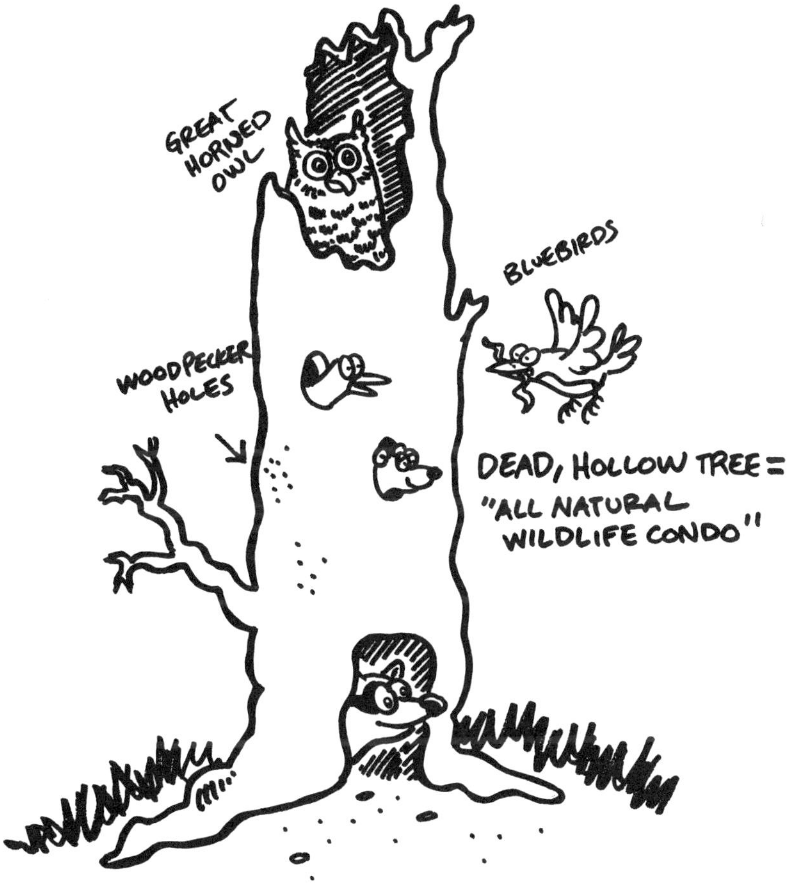
GREAT HORNED OWL
BLUEBIRDS
WOODPECKER HOLES
DEAD, HOLLOW TREE =
"ALL NATURAL WILDLIFE CONDO"

↑ Hole-nesting species will be attracted to birdhouses or trees with hollows, so consider leaving a decaying tree in your yard.

- Plant trees and shrubs to give birds and animals shelter from weather and predators. Chipmunks and other small animals will take cover in a stone wall.

If you're really serious about creating a suburban wildlife refuge, you can get your yard certified as an official Backyard Wildlife Habitat by the National Wildlife Federation (see Appendix).

Conservation in the City

If you live in the city you can still get out there and make conservation work! ↑ Plant flowers and vegetables in window boxes. Set out some bird feeders and enjoy bringing nature to your windowside. You'll probably be surprised at how many parks there are in your city and how few you've visited. Make a point of frequenting those pockets of countryside and follow a few rules:

↑ Don't feed the animals in the park. These animals may have become used to human handouts, but in the long run they'll be better off. Feeding keeps their numbers artificially high and allows weaker strains to survive, perpetuating an unnatural cycle of human dependence. If you insist on feeding urban wildlife, give them cracked corn and commercial duck food instead of human junk food.
• If there are established pathways, stay on them to avoid trampling plants.
• Pick up trash left by others. Recycle cans, bottles, etc.
• Visit a local zoo or aquarium. Many play an important role in breeding endangered species and educating the public about them. Zoos and aquariums house forty animal species that are now extinct in the wild. You can also join of "friends of the zoo" organization to help preserve these important institutions.

Besides parks and zoos, you may find patches of green around your city in the form of urban gardens. Community-tended plots and urban gardening groups have become increasingly popular recently. It takes lots of work to begin an urban garden because you have to deal with a whole range of issues: land ownership, potential pollution on the site, construction, neighborhood organization, and many other details. The key is long-term land control—at least ten years. Many groups set up a legal trust or nonprofit organization to protect the land. For advice, contact the American Community Gardening Association (see Appendix).

THE CHARLES "BEFORE"   THE CHARLES "AFTER"

↑ Another way to enhance the city environment is to join or start a group to clean up and revitalize an urban river. When the Charles River Watershed Association was formed in 1965, Boston's Charles River was "an open sewer," according to the organization. They fought for stricter sewage treatment and discharge standards, helped close leaching landfills that were polluting the river, and organized scores of cleanups along its banks. Today the Charles is classified as swimmable from source to mouth. People eat fish they catch there, and on a nice summer day the river is dotted with canoes and sailboats. For information about saving rivers in your area, contact the River Network (see Appendix).

Finally, a couple of additional ideas for city dwellers.

↑ If you don't have room for trees near your home, plant houseplants to fight global warming and indoor air pollution. Good choices are aloe, English ivy, and spider plants.

• Find an abandoned railroad spur and set about converting it to a walking or bike path. More information is available from the Rails-to-Trails Conservancy (see Appendix). The Conservancy has helped local and regional groups turn more than 3,000 miles of track into nature trails, bike routes, bridle paths, and cross-country skiing lanes.

What the AMC is doing in the city

The AMC's Boston, New York/North Jersey, Washington, D.C.; and Worcester, Massachusetts chapters offer a wide range of outdoor activities for urbanites. Programs include nature walks, canoe and sailing trips and instruction, outdoor photography workshops, and bike excursions. Urban chapters also sponsor conservation and trails committees, allowing members to get involved in environmental advocacy and trail work. In 1991 the AMC began a new program, AMC Urban Trails. Among its projects is development of a system of "greenways" stretching from Boston to the Canadian border. The new program works in conjunction with the AMC's Youth Opportunities Program (YOP).

- ↑ YOP trains teachers and kids from city schools in backcountry skills and conservation. Following YOP training, groups can take part in AMC-led trail work projects, such as construction of the Bay Circuit Trail in Boston.
- The New York/North Jersey chapter of the AMC has organized Appie Outing for Urban Kids (AOK), which each summer leads hikes for hundreds of young New Yorkers. That chapter's Urban Trails Subcommittee is also largely responsible for developing Pelham Bay Park, an urban wilderness in the Bronx.

5 Conservation in the Backcountry

So far, we've talked about how to mimic nature by using resources wisely in and around our homes. What about when we actually try living a while in the natural world? Conservation is just as important when we go hiking, camping, or canoeing as when we're at home. In 1990, the AMC launched its "Conservation**works**" campaign with a series of posters featuring cartoon characters promoting Earth-friendly practices. The posters, which now appear throughout the AMC's huts and other facilities, became the basis for other Conservation**works** projects, including (of course!) the book you're holding now.

Through educational workshops and publications, the AMC encourages people to respect nature and live in harmony with it. AMC naturalists use the book *Soft Paths* by National Outdoor Leadership School (NOLS)* Senior Staff Instructor Bruce Hampton and U.S. Forest Service Research Biologist David Cole to spread the word on enjoying the wilderness wisely.

---

* NOLS is based in Lander, Wyoming. *Soft Paths* is published by Stackpole Books, Harrisburg, Pennsylvania. Copyright © 1988 by the National Outdoor Leadership School.

↑ You don't have to bring all the trappings of modern life into the woods to enjoy yourself.

Hiking

- First, stay on established hiking trails to protect fragile mountain and forest plants from trampling.
- Soil erosion, inadvertent widening of trails, and creation of new ones are the most common types of hiker-related damage. Problems are more likely when trails are wet. If you find yourself hiking a muddy trail, stick to the established path, even if it means sloshing through puddles (wearing a pair of gaiters will help to keep you dry). Place fallen branches across any "informal" trails you find.
- Select the right footwear. Heavy lug-soled boots are unnecessary unless you're hiking steep rocky terrain. In other environments—wet boggy areas, for instance—they can seriously damage native flora. Opt instead for lighter boots.

- Stay on established switchbacks and encourage others to do the same by throwing debris across any shortcuts you find.
- If you're traveling in backcountry where trails don't exist, walk on the most durable surfaces—rocks, sand, snow, dry meadows, grasses, and sedges. Avoid following the paths of other hikers since this will encourage the formation of a trail through otherwise undisturbed wilderness. If you're hiking with several people, spread out.

Camping

Ideally you want to leave your campsite cleaner and in a more natural-looking state than it was before you got there. Here's how:

- Select a campsite that's either never been used before or one that's been heavily used. Sites showing low or moderate use will eventually regenerate if left alone, but a heavily-used campsite won't suffer much from additional use.
- Even high-use campsites deserve respect. Set up your tent on thick forest litter and duff. If you can't find a space with thick duff, opt for one that's naturally devoid of it (rock, gravel, or sand) rather than a place that's been eroded by human use. Steer

clear of meadows and the edges of forests since these are often critical wildlife habitats. After your stay, leave a high-impact site clean and attractive so others will use it.
- If you are in pristine wilderness, pick a site where you believe no one has camped before and leave it looking the same. Otherwise, you risk creating a campground where none existed before. Clues to prior use of a site include trampled vegetation and charcoal from old fires.

The best pristine sites are ones on durable surfaces—rock outcrops, gravel bars, sandy beaches, or snow and ice—since then there's no risk of disturbing plants and leaving traces of your stay for others to see. Camping on soil that has no natural vegetation is the next best choice. If you must camp on vegetation, select dry grass. Grasses grow in mats or tufts and are hardier than forest greenery.

↓ After picking a site, wear sneakers or other light shoes around camp to prevent trampling. Take alternate paths to water to avoid making a trail, and bring along a large collapsible water container to reduce the number of trips you'll need to make.

Your goal upon leaving a pristine site is just the opposite as when leaving a high-impact one: you want to discourage more use by others. Disguise disturbed areas with duff or other natural materials. Replace rocks or other objects you've moved and rake up flattened grassy areas with your fingers or a fallen branch.

No matter where you camp, don't dig trenches or excavate shoulder depressions to make room for your tent. And don't pull up plants, break off tree boughs for bedding, or move embedded rocks.

Fires

Once a given on camping trips, campfires are now controversial. They hurt natural systems by causing forest fires, charring soils, and reducing forest floor litter needed for soil regeneration. Many campers now use compact cooking stoves. But toasting marshmallows over a small blue flame somehow lacks the romance of the traditional campfire. There's also a counterargument: that camp fuel got from the ground to your stove through a massive industrial system. Which does worse damage?

In the end, people will probably always make campfires. Here are a few guidelines for doing it right:

- Set campfires in high-use campsites. In low- or moderate-use sites they deteriorate an area that otherwise might regenerate. Fires also don't belong in alpine or tundra environments, since trees growing in these harsh, cold conditions need all the energy they can get from decomposing litter that would serve as firewood.
- For safety, build fires far away from dry grasses, trees, branches, and root systems. Never leave a fire unattended, and don't make one on a windy day or during a drought.
- Use an established fire ring. If a site has more than one ring, pick the one that's had the most use and is in the safest location. Dismantle and clean up the others so later campers won't use them.

↓ In pristine wilderness, build only mound fires or pit fires:
  —Mound fires are the best choice in vegetated areas, since they protect flora. Make one by spreading a layer of soil at least three inches thick over a flat rock and building your fire on top. The mound of soil should be at least twenty-four inches in diameter to contain the fire. Find soil without disturbing plants by searching near uprooted trees, sandy streambeds, and exposed areas near rocks and boulders. After the fire is out, return the soil to wherever you found it.
  —Pit fires are permissible only where your campsite has exposed soil or sand—otherwise you will damage the local vegetation. To make a pit fire, dig a hole several inches deep and wide enough to prevent the fire from spreading into nearby duff and litter. Build your fire in the pit and, when you're done, take care to replace all the original surface soil.

fig a: MOUND FIRE

fig. b: PIT FIRE

- In areas where rocks are scarce, or on river trips, make your fire in a metal fire pan.
- For firewood, gather one- to two-inch-thick sticks lying on the ground. Avoid rotten wood—it makes a lousy fire but plays an important role in the forest's natural recycling. Collect only as much wood as you'll need for a small fire and don't burn toilet paper, scraps of food, or plastic. Pack them out.
- Clean up your fire before quitting a campsite. Burn wood to ash and crush any remaining pieces of charcoal. Consider the fire out when you can sift the ash through your fingers. Scatter ash, powder, and any unburned firewood over a large area far from your campsite. Leave the fire ring in place if you used a high-impact site. At a pristine site, camouflage your fire by rinsing off rocks used for mound fires. Cover the area used for pit fires with soil and leaf litter.

Trash

The Appalachian Mountain Club's credo on trash in the wilderness is "carry in/carry out." That means you bring back everything you came with, leaving no trash in the woods. Some tips:

↑ Start by minimizing the amount of "future trash" you bring into the woods. Instead of taking along bulky boxes for cereals, rice, and noodles, pack these foods in plastic bags that can be reused as trash bags later. Better yet, buy your food in bulk, then pack it up, eliminating excess packaging altogether.

- Unwrap items such as candy and granola bars before your trip, and pack several in one plastic bag.
- Avoid taking in cans and bottles and you'll have less trash to carry out.
- Carefully proportion food to avoid leftovers. If you do have leftovers, pack them out with the rest of the trash. Scattering food on the ground will attract animals and make them associate humans with a free meal. It's bad for the natural balance of things.

↑ Smelly fish is the one exception to this rule. Throw fish remains at least 200 yards from campsites. In bear country, where safety is the paramount factor, it's all right to throw leftover fish back into mountain streams as long as the water is running high. If it's a low volume stream, scatter the fish on the ground at least a quarter mile from camp. And wash your hands—bears have a very good sense of smell!

Wastewater

Backcountry conservationists encounter another quandary when it comes to using soap. The best choice is not to use it, since the chemicals in soap can alter the pH balance of water and hurt aquatic creatures. However, if you do use soap, choose a phosphate-free brand.

- Before widely scattering water used to wash dishes, separate out food scraps large enough to be packed out with your other garbage. Again, in bear country, it's okay to deposit small amounts of cooking water directly into a stream with adequate flow. Otherwise, pour it into a hole and cover with soil or sand. Pouring excess cooking water into a sump hole is also the best policy in winter.

↑ Whether washing dishes or yourself, carry a container of water at least 200 feet from the river or lake where you got it, and wet, lather up, and rinse there. The 200-foot distance allows wastewater to filter through the soil and contaminants to break down before re-entering the water source.

All the same water-disposal rules apply when brushing your teeth. Try using baking soda instead of commercial toothpaste. It's biodegradable and safer for the environment.

Sanitation

What do you do when you're communing with nature and "nature calls?" Everyone has encountered this problem, starting back with the family picnic when Mom told you to just "go behind a bush." Her advice was pretty sound, but here are some details for the backcountry:

- In remote areas where discovery by others isn't likely, the most environmentally benign thing to do is deposit fecal matter on top of the ground at least 200 feet from water. The sun and air help break it down faster. Smearing feces with a stick or rock speeds the process even more.

- In popular hiking areas, you are better off scraping a "cat hole" several inches deep in the organic layer of soil where there are many decomposing microorganisms. Cover the cat hole with excavated soil after use.
- Since river trips allow easier carrying of heavy equipment, canoeists should take along the makings of a portable latrine. You'll need a waterproof surplus ammunition can, several large heavy-duty plastic garbage bags, chemical quicklime, and a toilet seat. Line the box with one or two bags and fold over the rim. Place the seat on top. Add quicklime before and after use to reduce odors and to slow down decomposition and production of methane gas, which could explode the bag once it's sealed. Cover your "toilet" with a spare bag between uses to keep out flies. Squeeze the air out of each day's bag, close it, store it in the box, and deposit appropriately once you're back in civilization. Don't use this system for urine. It will make the bags unwieldy and reduce quicklime's effectiveness.
- On a canoe or kayak trip don't urinate directly into the river. Find a nonvegetated spot, like a rock, a good distance from water and other campsites. In winter, other hikers and skiers will appreciate your efforts to camouflage "yellow snow."
- Choose a communal latrine as an absolute last resort. These user-dug trenches tend to concentrate the wastes of several people, create large areas of disturbed soil and vegetation, and are more likely to aid in the spread of disease than are individual cat holes.
- At parks or other recreation areas, use outhouses if they are provided. In most cases, park employees are responsible for maintaining them and making sure waste gets disposed of properly.

⬆ While you're trying to do the right thing, see if you can do without toilet paper. It's slow to decompose when left on the ground and poses forest fire dangers when burned. If you do use it, be prepared to pack it out for proper disposal later. If you're game, try "natural toilet paper" in the form of leaves, grass, moss, or snow. Feminine napkins and tampons should also be packed out, except in bear country, where safety considerations dictate burning them and packing out the charred, odorless remains.

Wildlife

↓ When you are out in the woods, remember that you're in someone else's home and mind your manners. You can enjoy wild animals without bothering them. Before your trip, find out what types of wildlife you might see. Learn where nest sites and watering and feeding areas are likely to be, and avoid them. It might be tempting to get as close as you can to a feeding doe and her fawn, but such harassment could cause the doe to run away, expending energy that she needs for survival—and she might even abandon her baby in the process. Keeping human contact with wildlife as low-key as possible is especially important in winter, when food is scarce and animals must conserve energy. Watch animals quietly from a distance and behind cover and camp out of sight of a spring or other watering site.

Get Involved!

You say you've committed the backcountry dos and don'ts to memory, you're recycling like mad, your own house and yard are models of conservation—and still, you want more? Become a volunteer! By working with others committed to cleaning up the planet, you'll gain lasting friendships and really feel like you're part of the marvelous web of life that connects everything.

↓ Conservation organizations such as the AMC couldn't do their work without volunteer assistance. The AMC's army of volunteers helps maintain 1,200 miles of trails each year, including 350 miles of the Appalachian Trail. Opportunities range from full-blown service trips—"working vacations" in national parks and natural areas in Alaska, Wyoming, and Maine—to weekend and daylong trail work projects.

↑ Preservation of the country's remaining wilderness is a crucial issue throughout the country, and many groups dedicated to saving open space need volunteers. The Northern Forest Lands issue in the Northeast is an example—it's an effort to preserve as open space 26 million acres in northern New York, Vermont, New Hampshire, and Maine. This land is threatened by volatile development pressures. The AMC and other groups are working with private owners, timber companies, government agencies, and others to arrive at a wise policy that protects the land and the communities of the northern forest.

• Research is another area for volunteer involvement. Work related to ozone, river protection, solar energy, rare flora and habitat protection, air pollution, acid rain, and other issues are all important. Many organizations support research into environmental matters (see Appendix).

Whether you're studying wildlife on the Amazon, constructing trails in the Rockies, or turning a compost pile in the backyard, your decision to let conservation work is helping to heal the planet. Remember the forest scene at the beginning of this book? It couldn't work efficiently without every strand of the web of life doing its part. It's the same principle here. The more you conserve and the more people you convert to the conservation ethic through your efforts, the stronger our web will be.

Appendix

Chapter 2
Motor Vehicle Manufacturers Association, 300 New Center Building, Detroit, MI, 48202 (automobile air conditioning).

Automobile Importers of America, Suite 1200, 1001 19th Street, Rosslyn, VA, 22209 (automobile air conditioning).

Aseptic Packaging Council, 1-800-277-8088 ("brick pack" recycling).

Chapter 3
Pacific Landings Ltd., 1208 S.W. 13th Street, Suite 200, Portland, OR, 97205 (motor oil recycling kits).

Scott Paper Do-It-Yourself Business, Scott Plaza, Philadelphia, PA, 19113 (motor oil recycling kits).

Chapter 4
National Wildlife Federation Backyard Wildlife Habitat Program, 1412 16th Street N.W., Washington, D.C., 20036-2266 (send a plan and $5 for backyard wildlife refuge certification).

American Community Gardening Association, c/o Dennis Rinehart, Ohio State University Extension Service, 3200 West 65, Room 216, Cleveland, OH, 44102 (urban gardens).

River Network, P.O. Box 8787, Portland, OR, 97207-8787 (river protection)

Rails-To-Trails Conservancy, 1400 16th Street N.W., Washington, D.C., 20036 (preserving former rail lines as trails).

Chapter 5
Earthwatch, 680 Mt. Auburn Street, P.O. Box 403, Watertown, MA, 02272 (environmental travel).

The Nature Conservancy, 1250 24th Street N.W., Washington, D.C., 20037 (environmental travel).

Massachusetts Aububon Society, South Great Road, Lincoln, MA, 01773 (environmental travel).

Sources

Chapter 1
Carolyn Travers, director, Research Dept., Plimoth Plantation, Plymouth, MA.
Nanepashemet, manager, Wampanoag Indian Program, Plimoth Plantation.
The *Boston Globe*—10/9/90, 11/27/90, 1/10/91, 4/7/91, 4/8/91, 4/11/91, 4/16/91, 4/23/91.
Freedom Capital Management Corp., Boston, MA.
Earth Day 1990 fact sheets, Stanford, CA.
50 Simple Things You Can Do to Save the Earth. The Earth Works Group, Earthworks Press: Berkeley, CA.
World Wildlife Fund, March/April 1991 newsletter.
Political Ecology Group, San Francisco, CA.
"From Poison to Prevention." Sanford Lewis and Marco Kaltofen, National Toxics Campaign Fund, August 1989.
The Amateur Naturalist by Gerald Durrell, Knopf: New York, NY.
Earth Day 1990 Action Guide, created for New England Earth Day 1990 by *New Age Journal.*

Chapter 2
Jackie Dingfelder, planner, Office of Planning, Maine Waste Management Agency.
Mary Mears, spokeswoman, U.S. Environmental Protection Agency, Washington, D.C..
Ruth Ann Turner, spokeswoman, Massachusetts Chapter of the American Automobile Association, Rockland, MA.
Harold Garabedian, air pollution specialist, Vermont Chapter of the American Automobile Association.
Maryellen Harn, spokeswoman, New England Electric System, Westboro, MA.
Noel Perrin, professor of environmental studies, Dartmouth College, Hanover, N.H.
Energy Efficient Home Appliances Tips. The Massachusetts Audubon Society and Massachusetts Electric Company.
Action Guide. Concerts for the Environment: Minneapolis, MN.
50 Simple Things You Can Do to Save the Earth.
Enviro Action. The National Wildlife Federation, March 1991.
Tips for Saving Energy. New England Electric System.
The *Boston Globe*—4/17/91, 4/24/91.
The *Boston Herald*—7/11/90.
"Diapers in the Waste Stream: A Review of Waste Management and Public Issues," by Carl Lehrburger, Beaudry Communications, Washington, D.C.
Dydee Service, Inc., Boston, MA.

The Green Consumer. John Elkington, Julia Haiks, and Joel Makower, Penguin: New York, NY.

Gardener's Supply newsletter #6. Gardener's Supply: Burlington, VT.

Tips for Environmental Shopping. The Boston Food Co-Op: Allston, MA.

Reduce, Reuse, Recycle Fact Sheet #3, MassRecycle, Worcester, MA.

Earth Day 1990 Action Guide.

Massachusetts Water Resources Authority.

Simple Things You Can Do to Save the Earth Tip-a-Day Calendar, 1991. Andrews and McMeel: Kansas City, MO.

Chapter 3

Massachusetts Department of Environmental Protection.

Earth Day 1990 fact sheets, Stanford, CA.

Field Guide to Source Reduction and Recycling for Individuals. Massachusetts Audubon Society: Lincoln, MA.

50 Simple Things You Can Do to Save the Earth.

The *Boston Globe*—4/10/91.

Southeast Paper Manufacturing Company, Marietta, GA.

Better Homes & Gardens. May 1991.

Reduce, Reuse, Recycle Fact Sheet #3, MassRecycle, Worcester, MA.

Denis Hayes, president, Green Seal, Washington, D.C.

Earth Day 1990 Action Guide.

Access: Outdoors, Spring 1991: Seattle, WA.

Chapter 4

Karen Pelto, environmental affairs coordinator, Charles River Watershed Association, Auburndale, MA.

Leroy Stoddard, executive director, Boston Urban Gardeners, Boston, MA.

Jim Cardoza, biologist, Massachusetts Division of Fisheries and Wildlife.

Massachusetts Audubon Society Environmental Helpline.

E—The Environmental Magazine. May/June 1991.

Healthy Lawns Without Chemicals. Massachusetts Audubon Society.

National Wildlife. June/July 1991.

Earth Day 1990 Action Guide.

Sanctuary. Massachusetts Audubon Society: Lexington, MA. April 1991.

Melrose Free Press—4/25/91.

50 Simple Things You Can Do to Save the Earth.

Birds in Your Backyard. Massachusetts Audubon Society and Massachusetts Nurserymen's Association.

Chapter 5

Soft Paths. Bruce Hampton and David Cole. Copyright by National Outdoor Leadership School: Lander, Wyoming. Published by Stackpole Books: Harrisburg, PA.

About the Author

LISA CAPONE is a freelance writer specializing in environmental topics. Her work has appeared in *The Christian Science Monitor, E: The Environmental Magazine, Sanctuary, New England Business*, and *Appalachia Bulletin*. She lives in Melrose, Massachusetts, with her husband, Will Condon, and their children Sally and Scott.

About the Illustrator

CADY GOLDFIELD is a naturalist illustrator and writer from Marblehead, Massachusetts. She has illustrated two previous books: *Shrimps, Lobsters, and Crabs: Their Fascinating Life History* by Dorothy E. Bliss (Columbia University Press) and *Professor Farnsworth's Explanations in Biology* by Frank H. Heppner (McGraw-Hill Publishing Company).

About the AMC

The Appalachian Mountain Club is where recreation and conservation meet. Our 50,000 members have joined the AMC to pursue their interests in hiking, canoeing, skiing, walking, rock climbing, bicycling, camping, kayaking, and backpacking, and—at the same time—to help safeguard the environment in which these activities are possible.

We invite you to join the Appalachian Mountain Club and share the benefits of membership. Every member receives *Appalachia Bulletin*, the membership magazine that, ten times a year, brings you news about environmental issues and AMC projects, plus listings of outdoor activities, workshops, excursions, and volunteer opportunities. Members also enjoy discounts on AMC books, maps, educational workshops, and guided hikes, as well as reduced fees at all AMC huts and lodges in Massachusetts and New Hampshire. To join today, call 617-523-0636; or write to: AMC, 5 Joy Street, Boston, Massachusetts 02108.

Since it was founded in 1876, the Club has been at the forefront of the environmental protection movement. By cofounding several of New England's leading environmental organizations, and working in coalition with these and many more groups, the AMC has influenced legislation and public opinion.

Volunteers in each chapter lead hundreds of outdoor activities and excursions and offer introductory instruction in backcountry sports. The AMC Education Department offers members and the public a wide range of workshops, from introductory camping to the intensive Mountain Leadership School taught on the trails of the White Mountains.

The most recent efforts in the AMC conservation program include river protection, Northern Forest Lands policy, support for the American Heritage Trust, Sterling Forest (NY) preservation, and support for the Clean Air Act.

The AMC's research department focuses on the forces affecting the ecosystem, including ozone levels, acid rain and fog, climate change, rare flora and habitat protection, and air quality and visibility.

The AMC Volunteer Trails Program is active throughout the AMC's twelve chapters and maintains over 1,200 miles of trails, including 350 miles of the Appalachian Trail. Under the supervision of experienced leaders, hundreds of volunteers spend from one afternoon to two weeks working on trail projects.

The Club operates eight alpine huts in the White Mountains that provide shelter, bunks and blankets, and hearty meals for hikers. Pinkham Notch Visitor Center, at the foot of Mt. Washington, is base camp to the adventurous and the ideal location for individuals and families new to outdoor recreation. Comfortable bunkrooms, mountain hospitality, and home-cooked, family-style meals make Pinkham Notch Visitor Center a fun and affordable choice for lodging.

At the AMC headquarters in Boston, the bookstore and information center stock the entire line of AMC publications, as well as other trail and river guides, maps, reference materials, and the latest articles on conservation issues. Also available from the bookstore or by subscription is *Appalachia*, the country's oldest mountaineering and conservation journal.